BIRDS
EYE SPY

BLOOMSBURY
CHILDREN'S BOOKS
LONDON OXFORD NEW YORK NEW DELHI SYDNEY

BLOOMSBURY CHILDREN'S BOOKS
Bloomsbury Publishing Plc
50 Bedford Square, London, WC1B 3DP, UK
29 Earlsfort Terrace, Dublin 2, Ireland

BLOOMSBURY, BLOOMSBURY CHILDREN'S BOOKS and the Diana logo
are trademarks of Bloomsbury Publishing Plc

First published in Great Britain 2024 by Bloomsbury Publishing Plc
Text copyright © Catherine Brereton, 2024
Illustrations copyright © Xuan Le, 2024

Catherine Brereton and Xuan Le have asserted their rights under the Copyright,
Designs and Patents Act, 1988, to be identified as Author and Illustrator of this work

Published under licence from RSPB Sales Limited to raise awareness of the RSPB
(charity registration in England and Wales no. 207076 and Scotland no. SC037654).

For all licensed products sold by Bloomsbury Publishing Limited, Bloomsbury Publishing Limited will donate a minimum
of 2% from all sales to RSPB Sales Ltd, which gives all its distributable profits through Gift Aid to the RSPB.

Bloomsbury Publishing Plc does not have any control over, or responsibility for, any third-party
websites referred to or in this book. All internet addresses given in this book were correct at the
time of going to press. The author and publisher regret any inconvenience caused if addresses
have changed or sites have ceased to exist, but can accept no responsibility for any such changes.

A catalogue record for this book is available from the British Library

ISBN 978-1-5266-6288-0

2 4 6 8 10 9 7 5 3 1

Printed and bound in China by Leo Paper Products, Heshan, Guangdong

FSC
www.fsc.org

MIX
Paper | Supporting
responsible forestry
FSC® C020056

To find out more about our authors and
books visit www.bloomsbury.com
and sign up for our newsletters

For Bea – C.B.

To all lovely readers, may you delight in this book
and revel in the joy of the search game – X.L.

BIRDS
EYE SPY

Catherine Brereton

Illustrated by
Xuan Le

BLOOMSBURY
CHILDREN'S BOOKS

LONDON OXFORD NEW YORK NEW DELHI SYDNEY

Bird-spotting tip
Some birds might look different from each other depending on their position and whether or not they're flying. Look carefully at their features, including their beaks and markings, to find the right bird.

In the farmyard

It's the middle of summer in the English countryside. Swallows are flying to and fro. They fly to their nest, bringing food for their noisy, hungry young. Can you spy the little chicks? They're in a cosy nest, high up in the eaves of a barn.

All sorts of other birds are busy on this warm day on the farm, and other creatures too.

Can you find ...

12 swallows including 3 chicks

1 robin

1 blackbird

5 starlings

5 goldfinches

Beautiful butterflies are out and about at this time of year.

What else can you find in the scene?

A blackbird is singing from his perch. Look at his bright eyes and orange beak.

3 chickens

8 house sparrows

9 butterflies

5 cows

4 pigs

Above the fields

Can you spy the swallows here with their forked tails? They're flying over the fields on a warm day at the end of summer. Their shiny blue-black feathers shimmer in the sunshine. Watch them as they swirl and swoop, catching flies in mid-air.

Where are they off to?

What else can you see here?

The meadow is full of wild flowers such as poppies and daisies.

👓 Can you find ...

| **12** | **1** | **3** | **10** | **1** |
| swallows | buzzard | woodpigeons | jackdaws | grey partridge |

A buzzard is soaring in the sky, using its very good eyesight to watch for small animals down on the ground. Watch out, bunnies!

Look at the rabbits bounding across the field. Their tails bob up and down as they go.

4 rabbits

1 meadow pipit

6 cornflowers

7 poppies

8 ox-eye daisies

In the city

The swallows have started out on a very long journey. Off they go, flying through the city. Swallows are expert flyers, with bodies that are shaped for flying far and fast. Look at their long, pointed wings and forked tails.

What sights will they see on their journey?

What else is going on in the city below the birds?

Two young foxes are curled up, asleep in someone's garden.

Can you find ...

12 swallows

3 house martins

15 feral pigeons

5 starlings

2 foxes

House martins look like swallows, except they are black and white with a shorter tail. They often nest in towns.

Quack! Some ducks are paddling in the park pond. They are mallards, a common kind of duck.

1
sparrowhawk

1
lime tree

5
herring gulls

3
mallards

3
Canada geese

At the seaside

It's the beginning of autumn and the swallows have arrived at the seaside in the north of France. There are plenty of little flying insects for them to catch and eat. Some swallows are swooping right down, nearly touching the water, to catch them.

What can you see people doing on the beach?

What else can you spot?

Groups of grey seals rest on the rocks when they are not out hunting.

Turnstones get their name because they turn over pebbles to get at creepy crawlies underneath.

Can you find ...

| **12** swallows | **4** oystercatchers | **5** turnstones | **9** herring gulls | **2** common terns |

Herring gulls are large, clever, grey and white birds. Listen out for their loud, wailing cry.

1 sun star

2 grey seals

1 shore crab

3 starfish

4 limpets

Following a river

Watch our swallows as they journey through France. Each day they fly over new farms, riversides and villages. Each night they find a safe place to roost, or sleep, after their tiring journey.

Some deer are grazing near the edge of a wood. They are big males called stags. They use their large antlers for fighting.

Can you find ...

12
swallows

1
grey heron

1
otter

4
mallards

1
coot

Can you name some animals that like to live near rivers? Can you spot any of them in the scene?

There is a flash of black and white and shimmery green. It is a magpie flying across the fields.

On the riverbank a heron has pounced to catch a fish in its beak.

4
fallow deer

2
white storks

1
common wall lizard

6
collared doves

1
magpie

Red squirrels scamper up through the branches of the mountain trees hunting for autumn berries and nuts. Whoosh! They can climb fast.

A mother bear is teaching her cubs to find food.

Can you find ...

12
swallows

4
house martins

1
golden eagle

1
bearded vulture

13
European bee-eaters

Over the mountains

Now the swallows are high up in the mountains. On and on they fly. Swallows have frills of feathers above their eyes. These stop them from getting dazzled by bright sunlight – like sunglasses!

Do you like climbing up hills?

What can the swallows see beneath them?

Groups of brightly coloured bee-eaters are also making a long journey, from Spain down through Africa. They move in small flocks.

3
brown bears

2
ibex

3
red squirrels

3
red deer

16
pink thistles heads

A busy feeding point

The swallows have been on their journey for about seven days now. They have arrived at a place in the south of Spain which is very close to Africa. It's a resting and feeding point where they can prepare to cross the ocean. Tens of thousands of other birds have had the same idea! It's very busy and noisy!

Look, the flock of bee-eaters has travelled here too! Hello again!

What else can you see?

Can you find ...

12
swallows

7
black storks

7
white storks

10
common cranes

14
European bee-eaters

Can you find the child flying a kite?

Migrating cranes fly in large V-shaped flocks.

The white stork is a very large bird with a red beak and red legs. This place is marshy and is a brilliant place for storks to find food.

1	6	6	7	1
hoopoe	common terns	avocets	redshanks	honey-buzzard

The open sea

Join the swallows as they fly over the sea. It's only a few kilometres between Europe and Africa here, so the journey isn't long. It only takes them a few minutes.

Have you ever travelled over the sea? What else can you see in the water beneath the birds?

Long-finned pilot whales are really a type of large dolphin. They swim in groups called pods.

Can you find ...

12 swallows

6 long-finned pilot whales

12 flying fish

4 common dolphins

2 orcas

These black-headed gulls have lost their dark head colour ready for the winter. Now they have white heads with a dark spot.

A flying fish cannot really fly, but it leaps out of the water to escape bigger fish that are trying to catch it.

4
gannets

4
Manx shearwaters

1
great skua

4
black-backed gulls

7
black-headed gulls

Into Africa

Can you spy the swallows as they fly into Africa and journey through the tall mountains and low valleys of Morocco? They swoop down over rivers and lakes.

What else might they meet?

What can you see in the scene?

👀 **Can you find ...**

12 swallows

2 golden eagles

7 glossy ibises

2 mistle thrushes

1 green woodpecker

This eagle likes to hunt on its own. It has a hooked beak and powerful talons. Small birds need to beware!

Cactuses are spiky plants that can grow in dry places. They store water in their squashy stems.

A lizard is soaking up the midday sunshine.

1
crested lark

5
Moroccan rock lizards

11
palm trees

2
olive trees

1
cedar tree

The Sahara Desert

The swallows have made it to the sandy Sahara Desert. It is very, very hot here, so the swallows need to cross the desert at dusk after it has cooled down a little.

Do you think they'll make it?

What else can you see in the desert?

An eagle owl has woken up, ready to hunt bats, sleeping birds and insects that fly at night.

Can you find ...

12
swallows

5
camels

6
gazelles

1
ostrich

1
eagle owl

Ostriches have thick eyelashes that protect their eyes in a sandstorm. They are the biggest birds in the world.

The fennec fox has enormous ears! They let heat escape and keep the animal cool.

1
nightjar

8
house sparrows

1
cobra

6
scorpions

1
fennec fox

Into the rainforest

The swallows have made it to a rainforest in Africa. It is warm, wet and green there. The air is full of juicy insects for them to eat and the thick forest is alive with animals big and small.

Where will the swallows perch?

A gorilla is munching on bamboo stalks and leaves. She carries her baby on her back.

Grey parrots are very friendly and love company. Look at their reddish tails.

Can you find ...

12 swallows

1 chimpanzee

1 dwarf kingfisher

4 grey parrots

2 tarantulas

The shy okapi looks like a cross between a giraffe and a zebra! It has a very long tongue for tearing leaves from trees. Slurp!

What else can you find in the rainforest?

1 okapi

1 forest elephant

2 gorillas

1 pangolin

1 crocodile

The savanna

It's December in southern Africa. Can you spy the swallows as they swoop and dive? They're gliding over a grassy landscape called the savanna. Down in the grass, big animals are keeping an eye out for food or for danger in the hot midday sun.

Who is trying to stay cool in the heat?

Cheetahs are chasing antelopes. Look how fast they are all running.

A leopard hides up in a tree, waiting to pounce on its prey.

Can you find ...

12
swallows

2
lions

1
leopard

5
African elephants

7
zebras

Zebras like to stick together in groups. They eat lots of grass. Look at their stripes.

What other animals can you find in the grassland?

2 hippos

8 wildebeest

6 giraffes

1 Cape vulture

1 go-away bird

A new wetland home

At last! The swallows have arrived at their home for the next few months – miles and miles of flat mud and reedbeds. There are millions of tasty insects for them to catch and thousands of other swallows for them to make friends with. Each night they all sleep in a safe hiding-place in the reeds.

What a long, exciting journey!

What else lives here?

Weaver birds make beautiful basket-shaped nests out of reeds or grass.

👓 **Can you find ...**

12
swallows

5
antelopes

12
greater flamingos

5
great white pelicans

1
African fish eagle

The spoonbill gets its name because its beak is like a great big spoon. It uses it to scoop up creatures to eat.

Flamingos sometimes go to sleep standing on one leg. Could you do that?

4
African spoonbills

10
Cape teal

1
African jacana

2
African oystercatchers

2
weaver birds

The swallows' long journey

The swallows' amazing journey is called a migration. Every year these little birds spend the spring and summer in one place, where they make a nest and raise their young. When summer ends, they fly thousands of kilometres to another home, where it is warmer and there is plenty of food to last them until the next spring.

In this book we've followed a group of swallows that take a particular route. They spend summer in the UK or Ireland and make their journey southwards through France, Spain, cross into Africa and travel down through lots of countries in the centre of Africa. They pause in the rainforest in the Congo Basin, and they end up in South Africa. You can see this route in pink on the map.

Millions of other swallows from other parts of the world make similar journeys.

These little birds face many dangers on their journey such as bad weather, predators, lack of resting places and tall buildings they might collide with. Organisations like the RSPB provide safe spots for birds to rest along their way.

START

United Kingdom

France

Spain

Morocco

The whole journey takes around six weeks and is over 9,500 kilometres long. That's like you getting in the car and driving fast for three-and-a-half days without stopping!

We don't completely understand how swallows find their way. We think they use the position of the sun in the sky, which changes throughout the day, and that they have a kind of compass built into their brain. They probably also take notice of landmarks such as rivers, coastlines and mountains, and maybe they can tell good places to stop by smell.

THE SAHARA DESERT

CONGO BASIN

South Africa

FINISH

On the journey, the swallows usually fly in small groups of dozens of birds. They fly fast. Sometimes they might travel 50 kilometres in one day or sometimes they might manage 400 kilometres!

In late January, the swallows start to fly back again. They arrive in the UK in March and April. Often they will return to exactly the same nesting place year after year. Is there somewhere you return to every year?

Swallows are just one of the many animals around the world that migrate. Whales, reindeer, wildebeest, sea turtles and many butterflies and other birds do, too.

What amazing little birds swallows are! Will you look out for them arriving every spring?

What did you spot?

What a long, interesting journey the swallows make!
And what a lot of other animals they see
on their way.

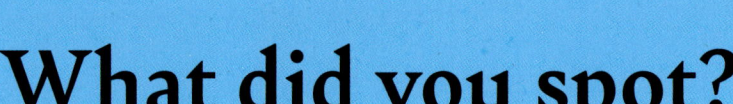

☛ Which place would you most like to visit?

☛ What was your favourite bird in the book?

☛ Can you name five birds or other animals from the
book without looking back?

☛ Have you seen any of the animals in this book in real life?

How many times did you see a little girl
flying a kite through the book? There are
8 to spot – why don't you go back and
count them all?